MAKING
SNEAKERS

BRUCE McMILLAN

Houghton Mifflin Company Boston 1980

For the sneaker makers of Lumberton

Library of Congress Cataloging in Publication Data

McMillan, Bruce.
  Making sneakers.

  SUMMARY: Photographs and text explain the steps in
manufacturing running shoes.
  1. Sneakers – Juvenile literature. [1. Shoes and
boots. 2. Rubber industry and trade] I. Title.
TS1910.M32   685′.3103   79-26069
ISBN 0-395-29161-5

How is this pair of sneakers made?

First, sheets of rubber and chemicals are heated and mixed between two huge rollers. They are mixed and mixed again. The sheets are hung up to dry, then fed into a heated machine. The rubber comes out of the other end of this machine in a long, smooth strip three inches wide.

The rubber is cooled with jets of water and rolled through powder to keep it from sticking. Then it is fed into a cutting machine that chops the long strip of rubber into pieces.

Now each rubber slab is set between two hot, sole-shaped metal molds. The molds push together with great pressure, squeezing the rubber into the shape of a sole. It stays in the heated mold for a few seconds, curing, so that it will keep its new sole shape.

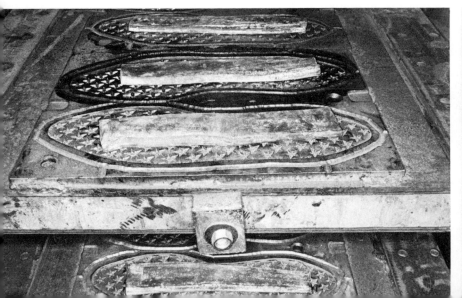

The rubber is peeled out of the molds. What were only slabs of rubber are now the bottoms of sneakers. With a die (a metal pattern that is as sharp as a knife), the cutter trims the soles using a machine that pushes the die down.

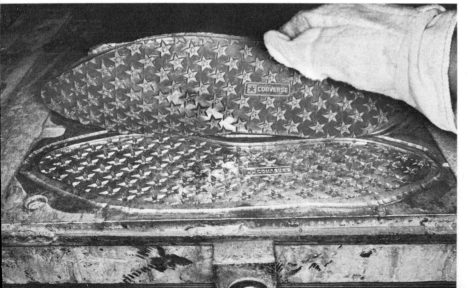
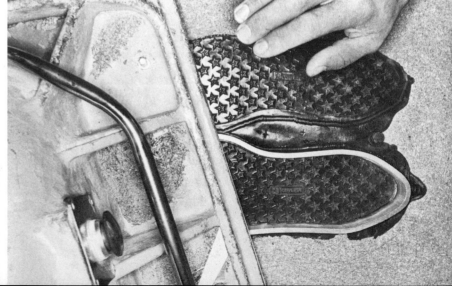

The rubber sole piece is glued onto another part of the sole, which has already been cut out with a die. This piece is made of a rubber-like substance and has microscopic air cells in it. The air cells will make the ground you run or walk on seem softer.

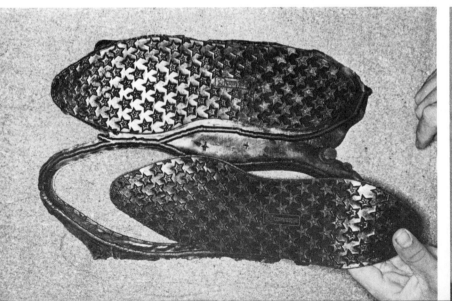
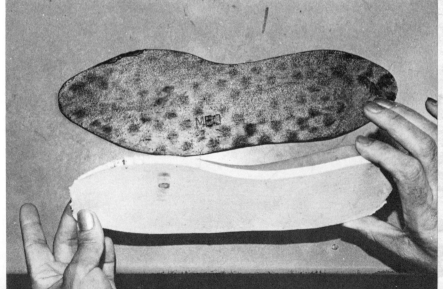

After the glue is dry, the sole is set into a specially designed sanding machine. This machine holds the sole tightly while rotating it around a sander, which trims and bevels the sides.

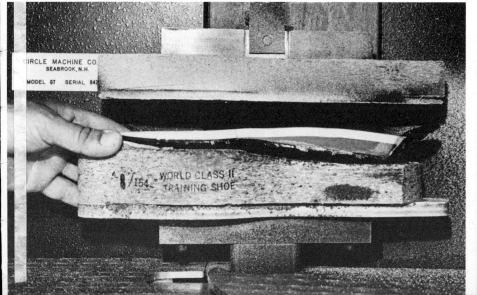

The finishing touches are done by hand on another sander, and then the soles are completed.

To make the top part of the sneaker, you start with a giant roll of fabric.

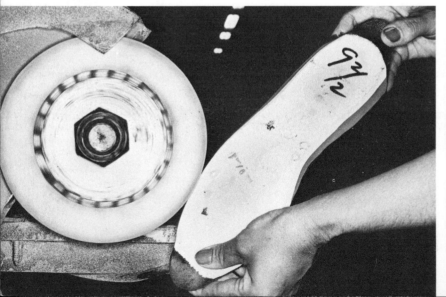

The fabric is rolled out, cut, and piled into layers. The cutter sets another sharp metal die (this one in the shape of the sides of the sneakers) into place to cut the layers of fabric.

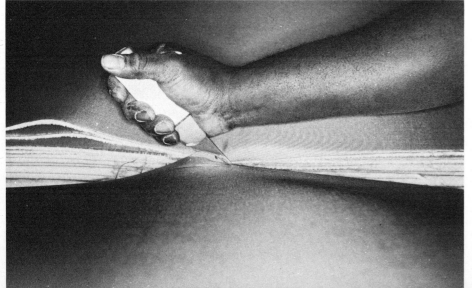

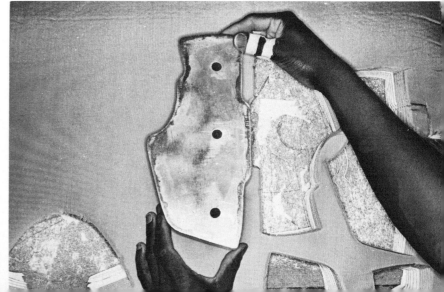

He swings a heavy press across. At the push of a button, the fabric is punched out with a die. This is also the way leather and vinyl are cut to make the other parts of the sneaker, such as the tongue, the toe, or the heel.

The stitcher takes the right and left sides of several sneakers, and matches them up. She sews them together up the back, one after another, and then snips them apart.

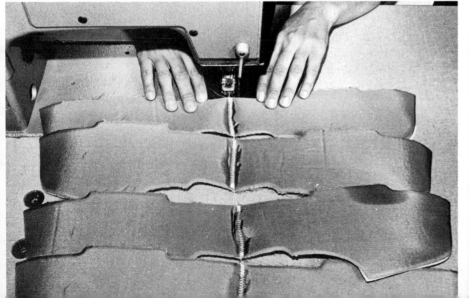

She stitches on some decorations with a machine that sews straight and stitches on stars with a machine that sews in the shape of a star.

Another stitcher sews a leather tab onto the back of the sneaker. Then she sews on a cushioned white collar. Once this collar is turned inside out, the part that will be around your ankle is finished.

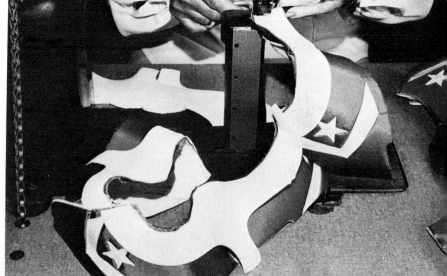

The stitcher sews on leather strips where the shoelaces will go. Then the sneaker is set into a hole-punching machine.

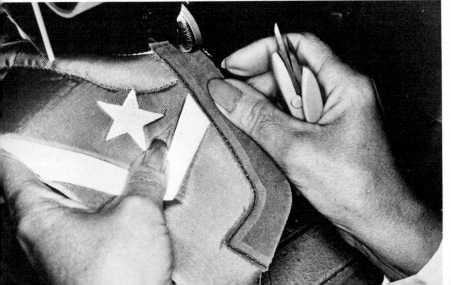

The machine punches the holes for the laces to go through.
To add extra strength to the heel, another stitcher sews on a big leather piece with double stitches.

Now the finished tongue is sewn onto the finished sides, first onto one side, then onto the other.

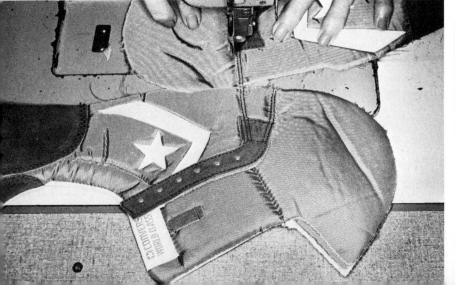

The last part to be sewn on is the leather toe piece. Then the sneaker is hung on a rack and is ready for the next step.

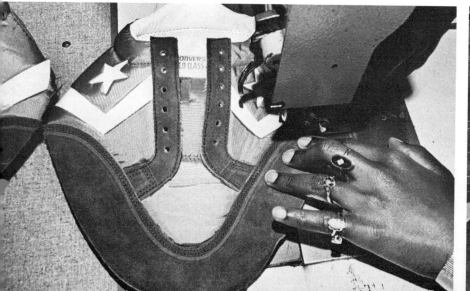

A heat-sensitive plate is slipped inside the heel, and the sneaker is set into a machine that clamps itself around the heel, holding it tight.

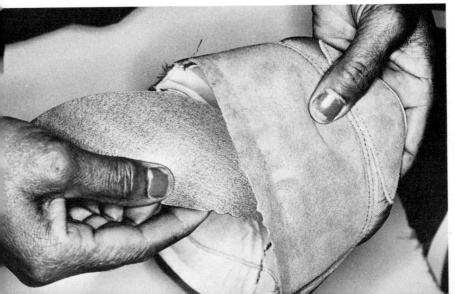
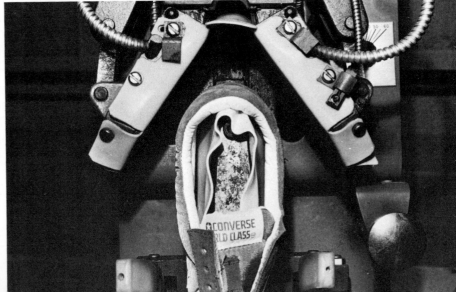

This machine, using heat and pressure, changes the shape of the plate inside the sneaker so that the heel stays in the shape of a heel.

Then the top part of the sneaker is set in place on a sneaker form, the same shape as your foot, with a fiber board at the bottom.

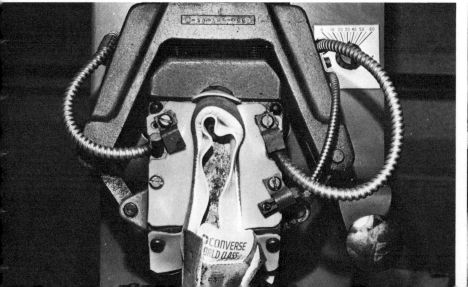
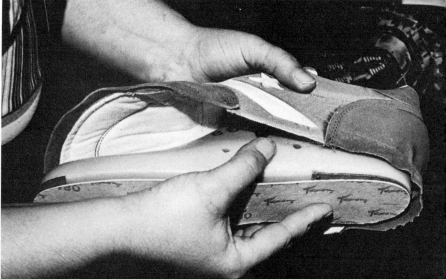

The whole thing goes toe first into a machine that holds the leather toe. The toe is pulled tightly into shape and glued with hot glue (about 300° F) to the fiber board underneath.

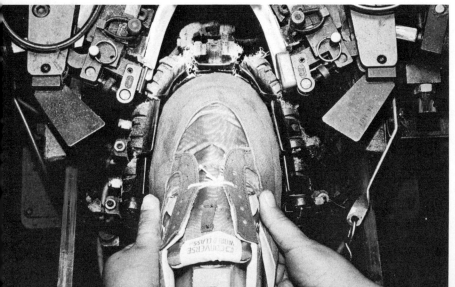

On another machine, the sides and heel are pulled into place with fingerlike clamps and glued to the bottom board. After the glue dries, the bottoms are sanded smooth.

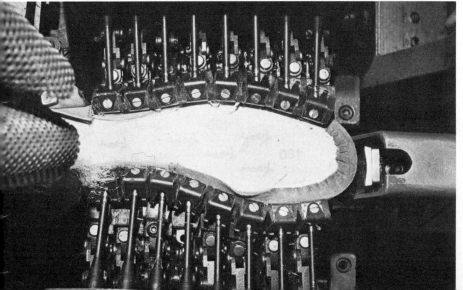
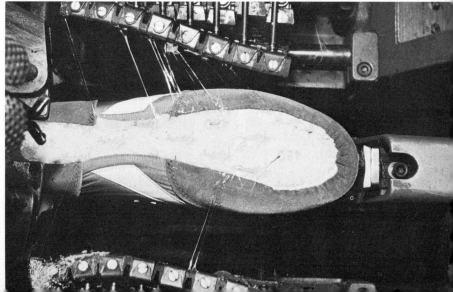

More glue (not hot this time) is painted on, a dab for the toe and a dab for the heel. The top is almost ready to meet the sole.

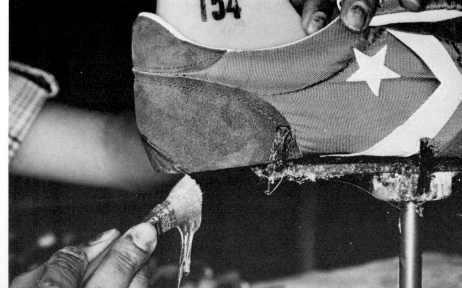

A wheel turns the top part of the sneaker around, and a machine brush spreads glue on the outside-bottom edges. Finally, with a hand brush, the middle is also painted with glue.

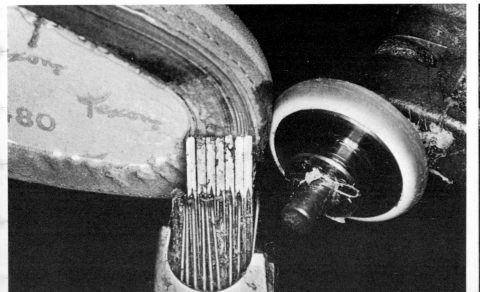

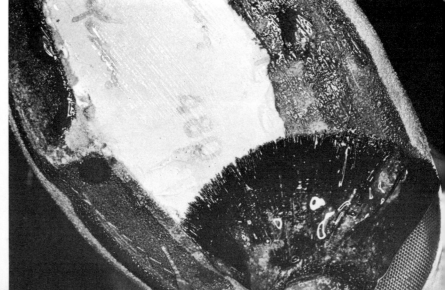

More glue is applied to the matching sole. The sole and top are dried, and then the glue is activated with heat.
The top and bottom are lined up and put into a press.

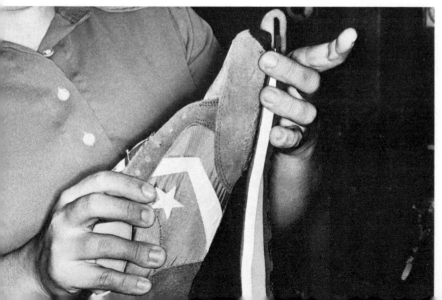

The press holds the sneaker tightly for a few seconds to let the glue work. Then the sneaker is set into another machine that pulls it off the sneaker form with a bang.

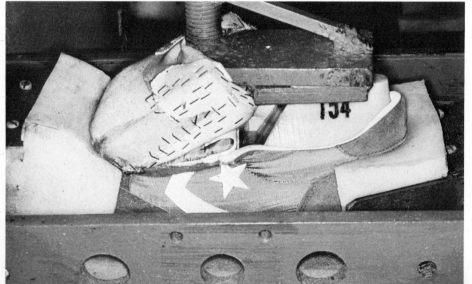
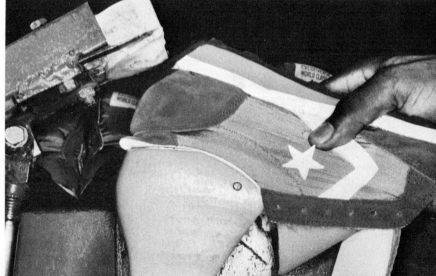

More glue is brushed on, but this time on the inside. It is spread all around, and then the cushioned insole is pressed in. There's only one thing left to do.

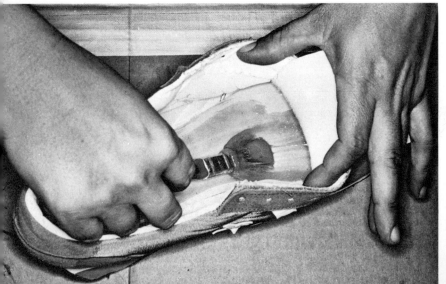
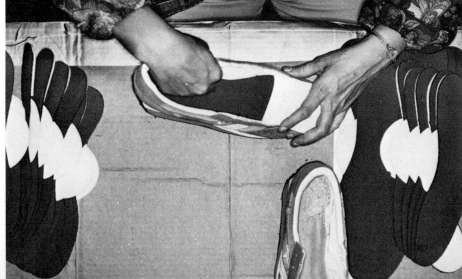

Shoelaces are tied on, and the sneakers are ready to go.

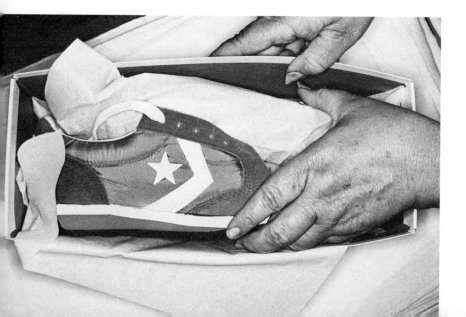

This book was photographed at the Converse Plant in Lumberton, North Carolina. I thank Converse, an Eltra Company, for their assistance in turning an idea into a book. I especially thank Billy Lyon, Pat Grooms, and Dan Floyd of Converse.

I also thank Mary Peterson's fourth-grade class of 1979, Shapleigh, Maine, for helping me decide what kind of sneaker I should photograph.

AUTHOR'S NOTE: The term **sneaker,** first coined in 1873, today encompasses all kinds of rubber-soled footwear, each with a specialized use. The type of sneaker in this book, also called a running shoe or jogger, reflects the changing times and is the most common type of sneaker worn today.